THE *NEW* SCIENCE OF SCIENTIFIC WRITING

USING READER EXPECTATIONS TO TRANSFORM SCIENTIFIC COMMUNICATION

GEORGE DAVID GOPEN

FHB | FIRST HILL BOOKS

FIRST HILL BOOKS
An imprint of Anthem Press

This edition first published in UK and USA 2026
by FIRST HILL BOOKS
75–76 Blackfriars Road, London SE1 8HA, UK
or PO Box 9779, London SW19 7ZG, UK
and
244 Madison Ave #116, New York, NY 10016, USA

British Library Cataloguing-in-Publication Data
A catalogue record for this book is available from the British Library.

Library of Congress Cataloging-in-Publication Data: 2026939299

ISBN-13: 978-1-80136-093-7 (Pbk)
ISBN-10: 1-80136-093-6 (Pbk)

Cover credit: Raya Kridel

This title is also available as an eBook.

IF THE READER IS TO GRASP WHAT THE WRITER MEANS, THE WRITER MUST UNDERSTAND WHAT THE READER NEEDS

An earlier version of this essay, coauthored by George Gopen and Judith Swan, was published in the November–December 1990 issue of *American Scientist*. That essay has been revised and significantly expanded by George Gopen in 2025.

TABLE OF CONTENTS

ACKNOWLEDGMENTS

I have many people to thank as I embark on a six-book series of looking at the writing of English from a perspective I call the Reader Expectation Approach to the English Language (REA). I started on this journey in 1978, when I was gifted (thanks to my neighbor Daniel Weinberg of Chicago's Abraham Lincoln Bookstore) with my first consulting job. I was asked by the Chicago law firm Epton, Mullin, Miller, and Druth to improve the writing of one of their young lawyers. With my knowing almost nothing about teaching writing, and with necessity being the mother of invention (I needed the extra funds to pay for a marital commute between Chicago and Philadelphia), I decided to work only on what struck my ear as most deficient in the young man's prose — his music, or the lack thereof. It worked. That was the beginning of my long development of analyzing what I call the music of English prose.

My most sincere thanks have to go to my friends, neighbors, and business partners, Barbara Croft Porter and her husband Jerry. They discovered my work when they moved to Durham in 2013 and decided to dedicate much of every day's energy and a great deal of money along the way in the attempt to make these ideas known worldwide. Without their care and their backing, this REA series of books simply could not have come to be.

I owe intellectual thanks to my partners in the consulting firm we called Clearlines. Joseph Williams, Gregory Colomb, Frank Kinahan, and I worked together from 1980 to 1990, with the booking genius of Joel Henning & Associates, which contracted for us many of the nation's top law firms and leading corporate legal departments. Joe Williams supplied the ideas, which were a practical application of concepts developed by the Prague School of Linguistics. For three years and more, these brilliant lawyers, our clients, poked holes in everything we had to offer. As we developed defenses to these hole-pokings, we slowly made our advice more solid. Eventually, the lawyers stopped poking holes and started taking notes. We owed a great deal to all of them. We knew we were on to something.

But we did not know why our advice was good advice. I started to discover the whys when we got a job offer that would not accept my consulting colleagues, because none of them had a law degree. Having that degree, in addition to the Ph.D. in English, I did that job solo – every Thursday, for a year, in New York City, flying there each week from my home in Durham, North Carolina. That firm was the second leading publisher of law and legal textbooks, Matthew Bender and Company. I am grateful to the Company for the opportunity, and especially to David Epstein for his friendship, and for the good administrative work he did in handling my 325 clients, in groups of 25 each. I did our four-day program — one day a week — 11 times. About the seventh time around, it suddenly occurred to me that everything I was telling them had something to do not so much with the language itself but rather with the language as experienced by readers. And so, in 1987, REA slowly started to take shape.

When I went solo, in 1990, I took on an assistant, Judith Swan, who wanted to trade in her career as a bench scientist for one in scientific writing. A brilliant mind, educated at Harvard and MIT, she learned my 16-hour presentation well enough over several years to become my partner and co-presenter. I was still doing this work only as a sideline — with the teaching of courses in Shakespeare, Music and Poetry, and Advanced Writing at Duke University as my "real" profession. Judy wanted to make scientific writing her full-time career. I felt if that was what she wanted, she needed a publication — for which she was not yet prepared. That is why I undertook the writing of the original "Science of Scientific Writing" article that appeared in American Scientist. Judy chose the scientific writing examples and made sure that our revisions stayed as close to the science as possible. She went her own way in 1999 and has had a distinguished career ever since.

I have included in this volume a short history — a remarkable history — of that article. In 2022–2023, I published two more long articles with American Scientist. I want to thank the journal, and especially its excellent editor, Fenella Saunders, for having such confidence in these new ideas.

I also feel a strong urge to thank my wonderful piano teacher, Miss Esther Hewes, with whom I studied from 1952 to 1963. She is the one who made me understand the concept of musical meaning and the practice of moving forward through time to an expected place of arrival. It took me decades to realize that same progression through time in music also applied to the reading of English prose. It all started with Miss Hewes.

I also wish to thank the Roxbury Latin School, the nation's oldest private school (founded in 1645), which for six very long years instilled in me the mental stamina and the power to concentrate that kept me working on the structure of prose for decades, until its essential movement quite suddenly became clear.

I also have another institution to thank, the Duke University Medical Center. Back in 1985, its Chancellor, Bill Anlyan, had enough interest in good writing to invite me to offer a 12-hour faculty workshop on scientific writing. I gave that course every year for 37 years, until interrupted by COVID. The Reader Expectation Approach to the English Language (REA) proved to be so popular that the 200 seats in the auditorium were spoken for, year after year, within 45 minutes after its email announcement was posted. I also need to thank the thousands of medical professionals who learned this material at Duke, many of whom brought me with them later on to the institutions which had lured them away.

I especially wish to thank Dr. De-Yin Jeng, who approached me several years ago with his offer to translate the original "Science of Scientific Writing" article into Mandarin. This modest beginning has flowered into this book being published in 18 languages. Sometimes good things just *happen* — especially when you are fortunate enough to be approached by good people.

Other individuals also move me to offer my thanks. For well more than a decade, I have benefited from the coaching skills of Jen Zobel Bieber. About half of everything I have written since her arrival in my life has been finished only so I would have something to send her before our next meeting. She has been a gem — perceptive, logical, and witty. I have learned much from her.

On a daily basis, I profit from the efforts both of my administrative assistant, the all-competent Julie Altschuler, and my agent, the ever-imaginative Robyn Wagoner.

Also on a daily basis, I have profited from the advice and support of my dear friend of 28 years, Dr. Laura Blanchard. As a superb pediatrician, she has done wonders taking care of the inner child in me.

I have also benefitted from the support of many individuals at my consulting institutions, but especially Victoria Seewaldt of The City of Hope, Stephen Bogdewic, then of the Indiana University Medical School, and Keith Sullivan of the Duke Medical School.

I owe profound thanks to my family for their love and support through everything along the way. That includes my sisters Emily Lipof and the late Sheila Schwartz, my son Alexander (an exemplary human being), and my "cousin" Bennett Lerner. They supply what no one else can.

Finally, I wish to thank Anthem Press, for their support in bringing out six volumes, in 18 languages, of an octogenarian academic. A pleasure to work with as colleagues, they excel in good judgment, astonishing speed, and the great care they take of the relationships they have with their authors.

DEDICATION

I dedicate this work, being published in
18 languages spanning the world, to my dear
friends, partners, and patrons,

Dr. Barbara Y. Porter and Jerry Porter.

They have spent much
of their energy in the past 13 years supporting
my work in every conceivable fashion. They
have done it with great good cheer, spirited
intellectual energy, and remarkable generosity.

Better friends there cannot be.

THE HISTORY OF "THE SCIENCE OF SCIENTIFIC WRITING"

"The Science of Scientific Writing" was published in 1990 by the esteemed journal *American Scientist*. The ideas were those of George D. Gopen, the founder and Director of the University Writing Program at Duke University, in Durham, North Carolina, USA. He was assisted by Microbiologist Dr. Judith A. Swan, who chose the scientific examples and took good care that their revisions were true – insofar as we could determine – to the intentions of the original passages.

The article immediately attracted attention, with offprints being requested in the first four months by scientists from 38 different countries.

In 2013, when *American Scientist* turned 100 years old, its editors selected this article as one of "The 36 Classic Articles" of the journal's first century.

The article was published again through the five editions of an anthology titled Exploring Animal Behavior: Readings from *American Scientist*, edited by Paul W. Sherman and John Alcock.

For more than 25 years, this article led *American Scientist's* citation index: That is, it had been cited in print by others more than any of the journal's other articles. As of 2026, it still remains fourth on that list.

According to Academia.edu, it is still being read some fifteen to twenty times a month by scientists from every continent.

The version you have here is a revision and expansion by Dr. Gopen of the original 8,200-word text into one of 13,400 words. It benefits from the additional knowledge he has acquired in the intervening 36 years. That is why it is now called "The *New* Science of Scientific Writing." It is scheduled to appear throughout 2026 in 18 different languages.

INTRODUCTION

Science is often hard to read. Most people assume that its difficulties are born out of necessity, out of the extreme complexity of scientific concepts, data, and analysis. I argue here that complexity of thought need not lead to impenetrability of expression; I demonstrate a number of rhetorical principles that can produce clarity in communication without oversimplifying scientific issues. The results are substantive, not merely cosmetic: Improving the quality of writing actually can improve the quality of thought.

The fundamental purpose of scientific discourse is not the mere presentation of information and thought, but rather its actual communication. It does not matter how pleased an author might be to have converted all the right data into grammatically correct sentences and paragraphs; it matters only whether a large majority of the reading audience accurately perceives what the author had in mind. In work begun by Joseph Williams, Gregory Colomb, and George Gopen in 1980 and developed by Gopen ever since then, we now have the ability to understand and predict the main interpretive acts of readers of the English language. From this has developed Gopen's Reader Expectation Approach to the English Language — which has come to be known as REA. It has helped tens of thousands of practicing scientists to succeed in obtaining grants and securing publications. This Introduction explains why our traditional writing pedagogy has failed us.

****

We have varied English writing instruction very little in the quarter millennium since we have been trying to teach it. I am here to put forward a revolutionary way of understanding, and therefore being able to control, the English language. I call it The Reader Expectation Approach to the English Language (REA).

Let us begin by understanding how we have hampered developments in teaching the language by not varying much from the way in which it was first conceived and published, in 18th century Scotland. Go back to that seminal effort, the Scottish philosopher Hugh Blair's *Lectures on Rhetoric and Belles Lettres* (1783): You will find him establishing the pedagogical modes that somehow got set in stone. There are two ways in which the lovely advice he offered has proved insufficient and has limited our efforts ever since then.

The first: In our teaching role, we make the nonsensical assumption that students reading the textbooks already can tell good prose from bad. Then all we have to do is display a bad sentence and come up with ways that allow the sentence to be improved, or even "corrected." After multiple examples like this, the students/readers then are supposed to be able to spot the difference between good and bad, allowing them to adhere to the good ever thereafter.

The second: Wherever possible, like Blair, we have cast writing instruction in the form of rules. Those rules describe the requirements of Society in its judgment of whether or not to have respect for you based solely on your writing. To prepare students to be able to enter into polite society, teachers need only memorize the rules, be able to spot when one has been broken, and then deduct points from the student's grade.

As a result of these two pedagogical habits, students have come to realize that they will fare well if they can function in ways that will not raise the ire or the red pen of the teacher. Good students do this well.

This produces a fake relationship between teacher and student: Those of us who got good grades in writing classes had learned the skills with which to please the teacher. But in the real world (that being the world you enter after the last time you graduate), the relationship between writer and reader becomes real, not fake: When you write as a professional person, you are presumed to be expert in that which you are saying; your job is then to make your readers become clearly aware of your thoughts. (This is the case whether you are proposing you be funded to advance nuclear science or are just announcing a committee meeting for Tuesday at 4:00.) The rhetorical term for this real world writer/reader rhetorical act is *communication*.

It is a lie or a fantasy to think that students in school are involved in the rhetorical act of communication. They do not think, having spent a brief time on line gathering information, that they have become the expert in that field. They do not conceive that their writing task job is now to fill the empty vial of teacher full of the milk of thought. Actually, they think about it in quite the opposite way: They think that teacher already knows just about everything student could find to say. Their rhetorical act of writing is therefore not *communication*; it is the narrower, more burdensome, far less interesting act of *demonstration*. If they can demonstrate to teacher that they have found a praiseworthy amount of information, and have presented it in error-free prose, then they will achieve a high grade. And if they seem to have improved in this effort since their last attempt, they will receive an improved grade.

This fake rhetorical relationship (a student writing for a fake audience, for an audience that has no need to know) does not function at all well in the real world. There, no one is likely to reward you just because you tried hard and filled all the pages with prose; nor are you likely to get credit for improvement. Can you imagine a funding agency like NIH looking at your grant application and declaring, "Your preliminary results are unconvincing, and we do not think you have the right personnel to carry this off: But we see you worked VERY HARD on this, so you get the million dollars." Or can you imagine a judge looking at a lawyer's brief and declaring, "This is a really poor brief, badly reasoned, and badly structured: But it is so much better than the one you turned in last month, you win the case." All those skills we learned in school in order to please teacher simply will not cut it in the real world. True communication cannot be attained just by avoiding error and following the old saws of advice like "To make it better, make it shorter," or "Avoid the passive."

Real communication has to be learned by figuring out not what teachers want but rather how readers go about reading. Knowing how readers go about the interpretive act will allow you to fashion your prose so that readers will actually understand your thought processes. Of course, for that to work, your thought processes also have to work — on a high level. The Reader Expectation Approach to the English Language, to which I introduce you in this volume, will equip you well for both the thinking and communicating processes.

In this Introduction, I want to cover the following ground: First, I want to take a closer look at the two ways in which I have called our writing instruction inadequate; then I want to explore why we have made so little progress in our teaching of writing in 250 years; and then I want to explore how all our writing instruction effort has been concentrating on the powers of the left brain, not the right brain.

Way #1 in Which Our Writing Instruction Has Proved Inadequate: A Nonsensical Assumption

In our teaching of English writing, we continually make the nonsensical assumption that students reading the textbooks or watching us in the classroom can already tell good prose from bad.

Besides pointing to Hugh Blair's *Lectures* — which I love to read — I will name only one other book, the beloved *Elements of Style,* by the endearing E.B. White, who was making use of the lectures of his mentor, William Strunk, Jr. This charming book reached its fourth edition in 1999, the first having been issued in 1918. It has never been out of print. It is the only writing text that is sure to elicit smiles to many in a lecture audience when mentioned by name. All the other myriad of writing texts that have appeared along the way tend to make this same nonsensical assumption. I have no interest in singling any one of them out by name.

I love re-reading Strunk and White: It is written with the lilt so representative of E.B. White in his children's books (like *Charlotte's Web*) or his many engaging essays. The man had a musical ear for prose. He had a quiet, eye-glint kind of humor. You feel he is in the room with you. But despite its wide adherence, this lovely book cannot teach you how to write well. It can act as a good example, which you might well try to emulate; but it does not teach you how to go about achieving that by yourself. In the 85 pages that constitute the Third Edition, the authors devote only a single, 19-page chapter to instructing us how to write. It is called "The Elementary Principles of Composition." I will comment here on just one of its 11 subcategories, entitled "Omit Needless Words."

Could there be a more sensible piece of advice? How could one argue against it? But hold on a moment: What can be the meaning of the word "needless"? Surely you must have felt there was some "need" for a word when you committed it to paper; so at least for that moment, that word to you did not seem "needless." If that word was truly "needless," then there is no "need" to retain it. Out it goes. But how do you go about knowing which of your once-needed words was actually "needless"? This advice is as old as Hugh Blair's 1883 Lectures. Here is what he says on pages 257–258 of Volume I:

> The first rule which I shall give, for promoting the Strength of a Sentence, is, to prune it of all redundant words. These may, sometimes, be consistent with a considerable degree both of Clearness and Unity; but they are always enfeebling. They make the Sentence move along tardy and encumbered.

Now to be sure, he is only addressing one sort of "needlessness" here, actual redundancy. I like the way he hedges in his second sentence. And I love the quality of the word "enfeebling." But in order to omit a needless word, the *writer* has to be able to perceive it is needless. In the publication process, it is usually the trained editor, armed with a set of pruning shears, that can spot the needlessicity of a given term. (Every once in a while, you have to invent a word for a new, specific purpose; and because it can be fun to do so, I do it, every once in a while.)

So to follow Strunk and White, you have to have an arms-length editor's ability to view your own prose in making this judgment of "needless." How can this be learned? No one seems to teach that. You just somehow have to know it. You have to already be one of those who is capable of being saved in order to be saved.

This whole blanket approach to needlessness is clothed in a religious kind of belief in right and wrong. There is a law that exists somewhere in the atmosphere; and our best critics can let us know about it, if we cannot see it for ourselves. However, I would counter-argue that there are rhetorical reasons that legitimize (to use Blair's example) redundancy. Take a sentence of mine from just two paragraphs back as an example.

> (Every once in a while, you have to invent a word for a new, specific purpose; and because it can be fun to do so, I do it, every once in a while.)

Whether you liked that sentence or not, let me show you a few rhetorical factors in play. First, it is encased in parentheses. That changes both its tone and its weight from the sentences that surround it. It is an aside. From that alone, the regular "rules" might function differently. It repeats itself, saying "every once in a while" twice. Surely one of these must be "needless," yes? No, not necessarily. The first of these gives context for the clause that follows it. The second repeats it as if to say, "Now I really mean this is the case only *every once in a while.*" It is as if I have overheard you criticizing that this shouldn't become a habit. My repetition says right back to you, "Yes, I know, I know; and I won't do it often." Its tone becomes grinningly sheepish. But there is more. That sentence begins and ends with that repetitive matter. Whenever I do that, it is quite possible I am constructing my favorite ancient rhetorical figure of speech — chiasmus. A chiasmus repeats two or more elements, but in reverse order. XYYX. The most famous 20th-century American chiasmus is John F. Kennedy's famous statement in his Inaugural Address:

> Ask not what your country can do for you; ask what you can do for your country.

Redundant words? Not at all. Why are they "needful"? Chiasmus often works by establishing a relationship between X and Y, then making a change in Y, which in turn makes a change in X. The reversal order of XYYX thus can signify a reversal in the sentence's substance.

Here is a translation of Kennedy's statement, perhaps the single most recognizable American political quotation since Lincoln's Gettysburg Address, 97 years earlier.

> Ask not what *your country* — the pork barrel that supplies opportunities for greed — can do for *you* — you greedy, self-consumed person you; ask what *you* — you good citizen in service of others — can do for *your country* — that transcendent virtue that lifts us all above ourselves.

The meanings of the Xs change because the meanings of the Ys have changed. Your country, the pork barrel, can become a transcendent virtue if you change yourself from a self-consumed person into one who is concerned for all those around you. It is no wonder that statement has become so memorable.

Can you imagine some 8th-grade teacher circling these words, commenting (in red) in the margin, "Redundant!", and taking off five points from the grade? I can.

Sometimes the addition of a seemingly needless word or two can make a positive contribution to the music of a sentence. (I will have much to say on this topic in two of the other books in this six-book series.) But here is an example of the opposite, which because of its musical insensitivity ruins a famous sentence:

> Do not ask what your country can do for you but instead what it is that you can do for your country.

The parallelism of the two halves suffers. I have spoiled its music.

Sometimes to keep a statement from being unmusical, the problem can be solved by the *addition* of a word or two. Since I've mentioned Lincoln's famous speech, why not turn to it for an example of how extending the number of syllables in a sentence can have a splendid musical effect? All Americans recognize the speech's opening phrase, and perhaps the whole first sentence:

> Four-score and seven years ago, our fathers brought forth on this continent a new nation, conceived in liberty and dedicated to the proposition that all men are created equal.

Had he written this for his high school English teacher (he actually had no high school education), teacher might well have circled "four-score and seven," commenting "Omit needless words," and suggesting as a replacement "Eighty-seven years," while taking off 5 points from the grade. But look at the ways in which the teacher would have damaged this famous opening gambit.

> Eighty-seven years ago, our fathers brought forth on this continent a new nation, conceived in liberty and dedicated to the proposition that all men are created equal.

"Eighty-seven years ago" has a clunky, unrhythmical sound to it, compared to the nice progression of "Fourscore and seven years ago." And the new, supposedly better, version no longer has a

Biblical ring to it. And the new version fails to balance nicely with the line that follows it, "our fathers brought forth on this continent," which has five stresses to it, the same as "Four-score and seven years ago" had. So there are many ways that what seems "needless" to a ruthless, syllable-cutting editor, thrilled with the virtue of reduction in size, can be meaningfully or artistically superior to its shorter, edited version.

To give you a bit more insight into how my Reader Expectation Approach addresses the problems inherent in trying to make prose clear and convincing, here is one more example, drawn from a popular writing textbook from 2003. It comes from a chapter entitled "Cut Wordiness."

> Speech itself is notoriously wordy, but one important difference between speech and writing is that writing offers the luxury of editing.

Do you find this sentence itself to be "wordy"? Out of 100 readers, *how many* of them have to vote for "wordy" in order for this sentence to be labeled "wordy"?

The real problem with this sentence has nothing to do either with the choice of words nor their number. The real problem here is structural. Neither taking five points off for wordiness nor the admonition "to make it better, make it shorter" is going to be of help in clearing up the underlying problem.

She [the author] starts off with a main clause ("Speech itself is notoriously wordy"); then she gives us another main clause ("one important difference between speech and writing is [X]"); and that "X" is itself a qualifying clause ("that writing offers the luxury of editing").

The real problem here is one that, to my knowledge, has never been noted in any book on writing: The sentence begins with a main clause but has no major mark of punctuation at the clause's end to create a Stress Position. A Stress Position is any location in a sentence (1) that brings the grammatical structure to a full halt, and (2) is marked as such by a properly used colon, semi-colon, or period. Beginning many sentences like this often causes what I have termed a "main clause first problem."

There is no reason you should be able to understand the import of the previous paragraph at this point, for several reasons: I have manufactured the term "main clause first problem"; very few people have ever heard of the term "Stress Position"; and no one of whom I know has ever tried to deal with explaining why a constant barrage of multiclause sentences like this one — where a main clause has no Stress Position — is highly problematic for a reader. I am not exaggerating when I say "constant barrage": According to my research, almost 50% of all sentences in scientific prose suffer from the "main clause first" problem. Even more shocking, in my last 16 years of analyzing the prose of thousands of scientists, I have not found a single person who is free from creating these problematic structures. "Wordiness" is not the problem: Heavily overladen and uninstructive structure is the problem. By the end of your reading *The New Science of Scientific Writing* you will understand

(1) that everyone has this problem;
(2) that no one to date has recognized the problem; and
(3) that the cure for it is straightforward.

Such problems! How can "Speech itself is notoriously wordy" possibly help a reader of her textbook who is not already aware of what this author is referring to as "wordy"? Speech is always made up of words. When do those words cross the line from efficient and communicative over to the penalty area of being "wordy"? I have faulted her sentence on the basis of its bad handling of clauses. That is a complex but entirely comprehensible matter; but would you understand anything about her sentence if I just labeled it "clausey"? To make progress in teaching and understanding how English functions, we must do better than mere name-calling.

Way #2 in Which Our Writing Instruction Has Always Come Up Short: Rules

Writing instruction was invented in 18th-century Scotland. Whom were they teaching? They were teaching children. Back then, in Scotland, children were meant to be seen and not heard. They were best taught in a digital fashion, by rules: Either something was right, or it was wrong. If the child was wrong often enough, the proper pedagogical method was to beat him. This was often done with a stick made of hickory or with a lashed switch. Such corporal punishment was continued in education in America well into the 20th century. Slowly, the stick or the switch was replaced by a ruler, which was applied to the erring student's hand. Eventually, the corporal punishment was replaced by the almost equally savage deduction of points from the grade. Teachers usually used a red pen for the deduction, perhaps in memory of the blood that used to flow from the hickory stick. Such a pedagogical approach can only be justified by the existence of rules, rules which must never be broken.

I was in my mid-70s before I finally understood the lyrics of a childhood song that we were taught in school at age 10 and sang joyfully — indeed, with great glee and happy gusto — on the playground at recess. Its lyrics began like this:

> School days, school days.
> Dear old golden rule days;
> Readin' and writin' and 'rithmetic,
> Taught to the tune of a hickory stick.

I knew all about the "rules" back then; but I had no idea what was meant by that allusion to the "hickory stick."

Teacher knew right from wrong. Teacher's mission was to get students to know right from wrong. Students knew their lives would be made immediately better if they could distinguish right from wrong for themselves. No wonder so many of us, though perhaps not actually beaten, remember our training in writing without great affection — indeed, often with distaste.

In the case of writing instruction, the concept of rules ruling the classroom is such a given that teachers could refer to it casually, as if it were a given piece of popular culture. Here is a typical textbook statement, which I found simply by opening the book to the middle and looking at the first thing that caught my eye:

> They wrote in their journals anything they wished, but for at least 30 minutes a day — and I collected, read, commented upon, and corrected them monthly.

The most problematic word here is "corrected." The teacher's assumption is that many things in those journals would be "wrong"; and that sufficient education would take place if the teacher simply "corrected" those errors.

To me, however, the rules have tended towards a kind of inhumanity, or anti-humanity. "It is what it is" — a declaration that always gives me problems — simply multiplies, concerning these rules, to "they are what they are." So often we wind up believing that whatever we are taught in school has always existed and must continue to exist, because "whatever is, is right." So often the justification for accepting the rules of the past without questioning them is simply "This is the way we always have done it." The older one gets, the more one tends to become protective of this mental acceptance of tradition.

When, in lecturing adult audiences on readers' expectations of paragraph structure, I state our schooling's concept that "each paragraph should begin with a topic sentence" is simply wrong in adult prose, I generate actual anger in the room: My pronouncement threatens to destroy something held dear, something assumed, something fundamental from their earliest education. I could get the same response by lecturing to an audience that believes in angels that there well may be no angels.

Adding a somber thought: The field of Psychology has taught us that if a parent badly abuses a child, that child, when grown to an adult, is highly likely to abuse its children. There is a clear parallel here in the way we teach writing: Since we were punished for not getting the rules right, we, in turn, [...].

Being somewhat limited in space here, I have chosen just a single rule to demonstrate the inadequacy of approaching such a complex and subtle question of meaning in language armed only with a hickory stick.

My choice of rules for this purpose is one that has modified itself slightly over time, in a mysterious way. The rule, acceded to by almost every textbook on composition, is the constant admonition to "Avoid the passive." Nowadays, it is common to encounter the rule slightly modified in the form of "usually avoid the passive," or "for the most part avoid the passive." Some books attempt to indicate a situation or two in which the passive can do a good job; but none of them attempt to tell us how often is too often. And almost none of them instructs us to distinguish between a necessary passive and an unnecessary one.

To explore this, let us return to Strunk and White's *Elements of Style.* In the third edition, point #14 in the chapter on principles of composition is headed "Use the Active Voice." We are cautioned, halfway through the section, that "This rule does not, of course, mean that the writer should entirely discard the passive voice, which is frequently convenient and sometimes necessary." But after noting a single example (a good example) in which the passive is to be preferred, they return to shooing the passive away.

The passive moves the furniture in a sentence around. As you will discover from the main essay in this volume, I argue that *where* a word shows up in the structure of a sentence will often determine *the use* to which that word will be put. If I am right about this, then the passive becomes a tool of major importance in controlling the English language: It gives you a tool with which you can make sure the furniture is all in the best possible places. The Reader Expectation Approach is the *feng shui* of linguistics.

"Jack loves Jill" is active; to make it passive, Jack and Jill change places — Almost all teachers of English will tell you that, compared to "Jill is loved by Jack," "Jack loves Jill" is the better sentence: It is shorter; and it is active, not passive. I would argue with this. To me, both of these sentences might be fine sentences; but they do different things. For most readers of English, the former, active sentence tells us the story of Jack; the latter, passive sentence tells us the story of Jill. These sentences should not be judged by active versus passive, nor by shorter versus longer; they should be evaluated on the basis of whose story is meant to be told. Punctuation and sentence structure, I argue, must be considered tools, not rules.

Here is this example in conversational action. Someone asks you "What about Jill?" If you begin with "Jack loves [...]," you might get interrupted with, "No, what about Jill?" You reply, "Be patient. I will get to that. Jack loves [...]," and your listener erupts in anger, "No! What about JILL?!" You begin again, this time with "Jill is loved [...],' and your listener is appeased. This is a reader expectation in English: Readers will read a sentence as being about whoever or whatever shows up as the sentence's grammatical subject. Very few textbooks ever mention anything like that.

Recently, I published some statements about the passive online, receiving a great deal of lively response. Almost everyone agreed that the passive was acceptable on occasion, but should not be overused; but no one had a definition of "overuse." So how much is too much? It turns out that frequency has nothing to do with the supposed problem.

Several years ago, I explored this question — how much is too much? — in the *Elements of Style*. I located the page on which they tell us "Use the Active Voice." I turned to the next page to explore whether they themselves had used the passive on that page. I found two examples. Then I read the following page, which produced two more. So I decided to read the entire 19 pages to find how many usages they had made of the much derided passive. They used it 38 times. That averaged two per page. Does this constitute "overuse"? It certainly struck me as nearly constant use.

But then I noticed that in these 19 pages, they quoted a great many examples of prose written by others; for statistical purposes, those foreign passages should not be counted as White's prose. So I decided to count the number of printed lines that were all his prose. That number of lines was 342. That means that with his 38 passive constructions, he averaged one use of the passive for every 9 lines that he wrote. My favorite example came from the very section that castigates the use of the passive. He tells us,

> Many a tame sentence of description or exposition *can be made* lively and emphatic by substituting a transitive [active] for some such perfunctory expression "there is" or "could be heard."

Is this an example of the annoying parental admonition "Do as I say, and not as I do"? E.B. White is a fine, even a glorious writer. He felt it necessary, in a book on how to write better, to use the passive once every 9 lines of his prose. Surely we need a better understanding of when the passive — or any other tool of writing — should be brought into use. Rules just don't cut it. Language is much too subtle, much too variegated for rules to rule its related pedagogy.

So I have come up with this single, seemingly dispositive statement about the use of the passive in English: The passive is not only acceptable, but it is actually better than the active in all those situations in which the passive does a better job than the active. There you have it. Now it just becomes necessary to be able to identify all those situations — a task the Reader Expectation Approach to the English Language will attempt to help you accomplish.

Why We Have Made so Little Progress in the Teaching of Writing Since its Principles Were First Put Forth in the 18th Century

I took on my first consulting client in 1978 — a lawyer who was smart but couldn't write. I needed the extra money to support a commuting marriage. If

I could not actually make this young man a better writer, I would be out of a job — and unable to see my wife. That explained to me why we have made so little progress in teaching composition.

Think of it: If you teach Freshman Writing and fail miserably, what is your punishment? You have to teach it again next term. If you teach the course and succeed brilliantly, what is your reward? You get to teach it again next term. There is no accountability. So naturally most of the concentration in the profession was on how to make the course less burdensome, more attractive, more fun, for both students and teachers.

I could not report to this young lawyer's law firm that he had attended classes, written a certain number of pages, and received a passing grade. I had actually to make him a better writer. None of the textbooks were going to help me to keep this job. Necessity being the mother of invention, and without having any idea where this would lead, I began laying the building blocks for REA just by looking with him at the music of his prose. It was all I knew to do. We looked at how the reader was (or was not) carried forward in his sentences by the rhythm of his prose. Ten months later, he was a better writer.

Left Brain versus Right Brain

Since the 1960s, scientists have long suggested that the left brain is geared towards logic, sequencing, linear thinking, facts, and mathematics; in contrast, the right brain is concerned more with imagination, intuition, rhythm, artistic qualities, nonverbal cues, and feelings. If you do not believe in this separation of brain functions, then consider all of the following just as a metaphor. It functions either way.

Almost all we teach in courses on writing and in our textbooks concerns the activities listed here as left brain activities. We teach students rules. We teach students organization. We teach students to expand their vocabularies. We especially teach students to avoid error. Students are urged to understand both what society expects of them as writers, and how to conform to those requirements as much as possible. We rarely teach anything in these courses that stems from right-brain activity — anything about sound, flow, or artistic balance. We almost never teach the relationship between the structure of prose and the flow of music.

Therein lies the problem: Human readers do not read like machines. Every sentence has a beginning and an end; and therefore every sentence must be read from left to right and through time, from its beginning until its end. The same is true for paragraphs. Depositing the information on the page has been taught as a left brain activity. You choose your facts; you make an outline form; you clothe those facts in recognizable sentences. If you selected enough impressive material, you will receive a good grade for your composition.

At no point in this process have we taught, until now, that readers have processes of their own, through which they can discover how to put your information together and turn it into thought. This process on the part of the reader engages the right brain extensively. We read sentences and paragraphs in much the same way that we listen to music. We search for "meaning" in both by attending to the journey from the beginning of every unit until its end.

Let me propose a definition for the word "meaning" in the context of music. Consider this. Let us say you are in a concert hall listening to a string quartet play a piece of Beethoven. To start the ninth bar of the second movement, all four players together play an emphatic chord. Freeze that moment. At that moment, each individual in that concert hall has a unique set of expectations as to what might come next. Consider Pat as a generalized example. Pat's expectations are based on a great many factors: Does Pat know this particular piece of music, and if so how well? If not, how well does Pat know other pieces of music something like this one? And how well does Pat know other pieces of Beethoven? But beyond that, Pat's expectations could be modified by how hot or how cold it is in the auditorium, or perhaps by what it was that Pat ate for dinner before the concert, or perhaps by whether or not during that dinner Pat had argued with the person sitting in the next seat. All of these influences, and more, combine to create Pat's unique set of expectations at that moment for what might come next. Now unfreeze the moment: All four string players play the next note, in a chord. I submit that the "meaning" of that next chord, for each individual listener, is the sum total of how that listener's unique set of expectations have been fulfilled and/or violated.

Meaning is a product of expectation at all times in music; I submit exactly the same holds true for the reader's experience in trying to make sense of prose. We take what we have been given, and the way in which it has been given, and combine that with what experience has taught us about how we are likely to get from this moment to the next moment. The "meaning" of that next moment in the reading process is the sum total of how all those expectations of ours become fulfilled and/or violated. Music and prose go about the act of meaning in essentially the same ways.

It then should make great sense that in order for us better to understand writing, in order for us to become better at doing it, we need to understand the ways in which we as readers are led from every individual starting point to every temporary or permanent conclusion. To understand this better, we must become more aware of the involvement of our right brain in the experience of reading.

Closure

Here, in one sentence, is REA's main pedagogical principle: To understand how English prose functions, we should stop looking at it from the perspective of what society demands, and instead look at it from the perspective of what readers actually do in the process of making sense out of prose.

Readers of English have relatively fixed *expectations* of where in the *structure* of the English sentence (or paragraph) to look for the arrival of certain kinds of *substance.* Put more briefly, readers of English know *where* to look for *what.*

There is no reason why you should be able to make great sense out of anything I have said in the previous two paragraphs. It is the job of the ensuing essay to bring you to a point of understanding how deeply different those two paragraphs are from all the other ways in which we have tried to explain the English language to the people who wish to gain more control of it.

If after reading the following essay, you wish to know more, I can promise you that it is my intention (and contractual agreement) to publish an entire book on the subject before the year 2026 has come to an end. Following that will be books on legal writing, punctuation, a text book, and a whole volume on elegance, concentrating on the rhetoric of Abraham Lincoln.

If you find this intriguing enough to follow up by reading more, and you wish to go to the trouble of revising your own writing habits to fit the expectations of English readers, I can promise you that

REA is capable of developing for you the following three wonderful capabilities:

— You will develop far greater control over your writing process, which will increase the quality of your product and (after a while) also increase the speed with which you write;

— You will come to have a far greater control over your readers' interpretive processes, so that they will get from your prose what you intend them to get from your prose; and

— You will find that REA principles will lead you back into your own thinking process, allowing you to recognize the ways in which your thinking requires further development.

I am always open to receive your questions, should you have any. I wish you joy in the journey.

CHAPTER 1

WRITING WITH THE READER IN MIND: EXPECTATION AND CONTEXT

Readers do not simply read; they interpret. Any piece of prose, no matter how short, may "mean" in ten (or more) different ways to ten different readers. This methodology of reader expectations is founded on the recognition that readers take the great majority of their clues for interpretation neither from the choice of nor the meanings of words but rather from the location of words within the structure of the English sentence. *Where* a word appears in a sentence will control to a great extent *the use* to which it will be put by a reader.

This interplay between substance and structure can be demonstrated by something as basic as a simple table. Let us say that in tracking the temperature of a liquid over a period of time, an investigator takes measurements every three minutes and records a list of temperatures. Those data could be presented by a number of written structures. Here are two possibilities:

a) t (time)=15′ T (temperature)=32°; $t = 0'$ $T = 25°$; $t = 6'$ $T = 29°$; $t = 3'$ $T = 27°$; $t = 12'$ $T = 32°$; $t = 9'$ $T = 31°$

b)

Time (minutes)	**Temperature (degrees Celsius)**
0	25
3	27
6	29
9	31
12	32
15	32

Precisely the same information appears in both formats; yet most readers find the second far easier to interpret. It may be that the very familiarity of the tabular structure makes it easier to use; but, more significantly, the table structure provides the reader with an easily perceived context (time) in which the significant piece of information (temperature) can be interpreted. The contextual material appears on the left in a pattern that produces an expectation of regularity — that of every 3 minutes; the interesting results appear on the right in a less obvious pattern, the discovery of which is the point of the table.

If the two sides of this simple table are reversed, it becomes much harder to read:

c)

Temperature (degrees Celsius)	**Time (minutes)**
25	0
27	3
29	6
31	9
32	12
32	15

Since we read from left to right in English, we prefer the context on the left, where it can more effectively familiarize the reader. We prefer the new, more important information on the right, since its job is to intrigue and educate the reader.

Information is interpreted more easily and more uniformly if it is placed where most readers expect to find it. These needs and expectations of readers affect the interpretation not only of tables and illustrations but also of prose itself. The most innovative and important discovery of REA can be stated simply: Readers of English have relatively fixed expectations about where in the structure of prose to look for the arrival of certain kinds of substance.

On the sentence level, what are those particular items of substance? They are the answers to four essential questions that a reader must correctly perceive if the reader is to make of the sentence precisely what the writer intended to convey:

1. What is going on?
2. Whose story is this?
3. How does this sentence link backward to its predecessor?
4. Which word or words should I be reading with extra emphasis because they are the stars of this show?

I propose that readers of English know where in the sentence to look for these answers. We all know these things instinctively as readers; REA helps us to become aware of these locations consciously as writers. Once a writer knows where to put these important pieces of information, the writer gains superior control over the reader's interpretive process. Because good writers (a small percentage of those who write) are intuitively aware of these expectations, their prose has what we call "shape." It is easy to read, always taking us forward, both within sentences and from sentence to sentence.

This underlying concept of Reader Expectation is perhaps most immediately evident at the level of the largest units of discourse. (A unit of discourse is defined as any progression of words that has a beginning and an end: a clause, a sentence, a paragraph, a section, an article, even a book — all units of discourse.) A research article, for example, is generally divided into recognizable sections, sometimes labeled Introduction, Experimental Methods, Results, and Discussion. When the sections are confused — when too much experimental detail is found in the Results section, or when discussion and results intermingle — readers are often equally confused. Even though smaller units of discourse do not often label their functional divisions, readers have definite expectations all the same: They search for certain information in particularly familiar places. If the expected information constantly arrives in the expected location, readers read with ease, with a justified conviction that they have perceived the writer's intended meanings. Conversely, if these structural expectations are continually violated, readers are forced to make too many interpretive decisions on their own, replacing a confidence of perception with an uneasiness of guessing. All too often, the reader is conscripted into the role of being the text's co-author. As the complexity of the content increases, even moderately, the possibility of misinterpretation or non-interpretation can increase dramatically.

I present here examples that demonstrate the results of applying REA to research texts in the scientific literature. Using several passages from research articles, I have suggested ways of rewriting them by applying principles derived from the study of reader expectations. I have not sought to transform the passages into "plain English" for the use of the general public; I have neither decreased the jargon nor diluted the science. I have striven not for simplification but for clarification.

Example 1:

The smallest of the URF's (URFA6L), a 207-nucleotide (nt) reading frame overlapping out of phase the NH_2-terminal portion of the adenosinetriphosphatase (ATPase) subunit 6 gene has been identified as the animal equivalent of the recently discovered yeast H^+-ATPase subunit 8 gene. The functional significance of the other URF's has been, on the contrary, elusive. Recently, however, immunoprecipitation experiments with antibodies to purified, rotenone-sensitive NADH-ubiquinone oxido-reductase [hereafter referred to as respiratory chain NADH dehydrogenase or complex I] from bovine heart, as well as enzyme fractionation studies, have indicated that six human URF's (that is, URF1, URF2, URF3, URF4, URF4L, and URF5, hereafter referred to as ND1, ND2, ND3, ND4, ND4L, and ND5) encode subunits of complex I. This is a large complex that also contains many subunits synthesized in the cytoplasm.

If you were to ask ten people why this paragraph is hard to read, nine are likely to mention the technical vocabulary; several will also suggest that it requires specialized background knowledge. Those problems turn out to be only a small part of the difficulty. Here is the passage again, with temporary substitutions for some of the most difficult words:

> The smallest of the URF's, an [A], has been identified as a [B] subunit 8 gene. The functional significance of the other URF's has been, on the contrary, elusive. Recently, however, [C] experiments, as well as [D] studies, have indicated that six human URF's [1-6] encode subunits of Complex I. This is a large complex that also contains many subunits synthesized in the cytoplasm.

Although it may now be easier to survive the journey through the prose, the passage is still difficult. Any number of questions present themselves: What has the first sentence of the passage to do with the last sentence? Does the third sentence contradict what we have been told in the second sentence? Is the functional significance of URF's still "elusive"? Will this passage lead us to further discussion about URF's, or about Complex I, or both?

Information is interpreted more easily and more uniformly if it is placed where most readers expect to find it

Knowing a little about the subject matter fails to clear up all the confusion. The intended audience of this passage would probably possess at least two items of essential technical information: First, "URF" stands for "Uninterrupted Reading Frame," which describes a segment of DNA organized in such a way that it could encode a protein, although no such protein product has yet been identified; second, both ATPase and NADH oxido-reductase are enzyme complexes central to energy metabolism. Although this information may provide some sense of comfort, it does little to answer the interpretive questions that need answering. It seems the reader is hindered by more than just the scientific jargon.

To get at the problem, we need to understand better how readers go about reading. Here is the first reader expectation to be considered — the only one not raised by the four questions above.

CHAPTER 2

SUBJECT–VERB SEPARATION

Look again at Example 1's first sentence. It is relatively long, 42 words; but the sheer number turns out not to be the main cause of its burdensome complexity. Long sentences need not be more difficult to read; they are only more difficult to write. Sentences of over 100 words can flow easily and persuasively toward their clearly demarcated destinations; but a seemingly simple but badly written ten-word sentence can completely confound readers without their even being aware of the problem. All well-wrought sentences, of any length whatever, have something in common: Their structure presents information to readers in the places where the readers both need and expect it.

The first sentence of Example 1 does just the opposite: It burdens and obstructs the reader because it violates a crucial modern-day English reader expectation: Today's readers expect that the moment we are handed a grammatical subject, we shall immediately thereafter encounter its verb. This is a relatively recent development. Two hundred years ago, great English prose writers copied Latin sentence structure by placing the subject at the beginning of the sentence and the verb at the end. We seem to have hung on to this Latinate construction when we try to write prose with the utmost of formality.

In our Example, the grammatical subject ("the smallest") is separated from its verb ("has been identified") by 23 words — more than half the sentence.

Because of our contemporary expectation of finding the verb immediately after the appearance of its subject, anything that intervenes between the subject and its verb is read as an interruption — as something of lesser importance. Think of this in terms of sound dynamics in the reader's ear. Once we are presented with a subject, we expect, we want, we need the verb to follow immediately. Subject–verb is a necessary combination: The verb is what the subject is doing; the subject is the doer of the action that is announced by the verb. For a reader to perceive what the writer is trying to say, the subject and its verb have to "speak" to each other at the same (silent) decibel level. In order for that to happen, any material that interrupts has to be read softer (and perhaps faster) in order to sustain the energy level of expectation accorded the arrival of the verb. By *locating* information between a subject and its verb, the writer is sending an *instruction* to the reader to consider that information of lesser importance than the rest of the sentence. The longer the interruption lasts, the more likely it becomes that the "interruptive" material actually contains important information; but its structural location will continue to brand it as merely interruptive. Unfortunately, the reader will not discover its true value until too late — until the sentence has ended. The damage has already been done. If this happens often, readers become increasingly fatigued — and therefore increasingly non-receptive. This one violation of reader expectations, if encountered often enough, can produce that foggy, cotton-headed sense of non-comprehension from which we all suffer when reading badly constructed prose.

This raises an important concept that you should keep in mind throughout this essay — a concept that I call "reader energy." Readers summon reader energy at the beginning of each unit of discourse — whether a clause, a sentence, a paragraph, or any larger unit — enough energy presumably required for the reading of that unit. Any time that unit requires from the reader more energy than had been initially summoned, the result can be a small amount of fatigue. When this happens continually, the fatigue becomes a real problem. It can also lead to confusion, even to the point of non-comprehension. The best writing never over-taxes reader energy.

In this first sentence of Example 1, the relative importance of the material that separates subject from verb is difficult to evaluate. The material might conceivably be quite significant, in which case the writer should have positioned it elsewhere to reveal that importance. Here is one way to accomplish this, by getting subject and verb together:

> The smallest of the URF's is URFA6L, a 207-nucleotide (nt) reading frame overlapping out of phase the NH_2-terminal portion of the adenosinetriphosphatase (ATPase) subunit 6 gene; it has been identified as the animal equivalent of the recently discovered yeast H^+-ATPase subunit 8 gene.

But what if that intervening material was intended by the writer to be a mere aside, one that diverts attention from more important ideas? In that case, the writer should have deleted it, allowing the prose to drive more directly toward its significant point:

> The smallest of the URF's (URFA6L) has been identified as the animal equivalent of the recently discovered yeast H^+-ATPase subunit 8 gene.

Only the author could tell us which of these revisions more accurately reflects her intentions.

CHAPTER 3

THE STRESS POSITION

These revisions lead us to a second major reader expectation: Readers of English feel the need to give the greatest emphasis in a sentence to any piece of information that arrives as the grammatical structure is coming to a full halt. This structural location is worthy of a name of its own — the Stress Position. From a 56-year career as a teacher and consultant on writing, I feel confident in saying that the violation of this one expectation is the single most widespread problem with writing the English language. As of 2026, having looked for this problem in the writing of tens of thousands of students and clients, I have only encountered four people who had been able to intuit the reader's need for the Stress Position and fulfill it constantly. Most people do it badly almost all of the time. If you can follow this one piece of advice alone — to place the information you most want stressed in a Stress Position — you can transform the quality of your writing. This can result in your grants being funded and your papers being published. It becomes necessary, then, to understand what a Stress Position is and how it functions.

In English — and especially in American English — I suggest that the Stress Position has its power because of the American value given to the act of closure. In America, we love to applaud loudly at the end of a well-performed piece of music; we love to celebrate when a business goal has been achieved; and we love the sense of great relief when some long task — like an extended journey — has successfully come to its conclusion. Similarly, while reading, we feel comfortable in conserving a major amount of our reader energy to be able to expend it at any moment when the grammatical structure of the sentence comes to a full halt. If we still have enough reader energy to stress the material located in a Stress Position, and if the writer has taken the trouble to place that stress-worthy material there, then the writer's intentions will be clearly communicated to the mind of the reader. If I am right about this concept of the Stress Position — and a lifetime of work convinces me that I am right about it — wouldn't it be foolish to place your most important information elsewhere in the sentence on a regular basis? And yet that is what an overwhelming majority of writers do.

So what is a Stress Position? A Stress Position is any moment of full syntactic closure. That is created by any properly used period, at the end of the sentence; but it is also created by any properly used colon or semi-colon in the middle of a sentence. For that use to be "proper," the clause that ends at the colon or semi-colon has to be capable of standing by itself as a full sentence. (We call such a clause a "main clause.") Why? Because only if it could stand by itself as a sentence could it offer the reader the opportunity to experience full syntactic closure. Since that clause could be a sentence all by itself, the colon or semi-colon is a powerful enough mark of punctuation to signal the reader to effect full syntactic closure.

A Stress Position can never be created by a comma. The comma is the only punctuation mark in the English language that does not announce its function at the moment of its arrival. Whenever you encounter a question mark, or an exclamation point, or a period, you are confident that you know what that mark is instructing you to do. But because there are so many different uses of the comma in English, you can never know what kind of comma it was trying to be until you have gone past it. It could signal the interruption of a single word, like "however"; it could signal the beginning of another main clause; it could signal that the first item on a list of items has just ended, and the rest of the list will follow immediately. We never know what kind of comma it is trying to be when we first encounter that comma; and, as such, it cannot offer us grammatical closure. It cannot create a Stress Position.

Since the colon and the semi-colon, mid-sentence, must always come at the end of a full main clause, these two marks are capable of producing a Stress Position mid-sentence. The colon is badly taught in American schools; and the semi-colon is rarely taught at all. Being able to construct a Stress Position in the middle of a sentence is such an important tool in the writing of sophisticated English about complex thought that it is worth our while to explore how these marks function.

A colon tells the reader to expect that the second half of the sentence will somehow exemplify the first half. Sometimes it will introduce a list of examples:

> In order to form this new company, we first have to be sure of three things: (1) The loan from the bank must be secured at 5% or lower; (2) our partners have to be willing to commit themselves before the end of June; and (3) our lawyers must be sure that we are not in violation of any of the relevant regulations.

In these cases of restatement, the material that follows the colon must be a main clause.

You will be helping your readers to distinguish between these two colon uses (for a list or for a restatement) if you capitalize the word that begins that second main clause. The capital letter instructs new readers to take in a breath of reader energy large enough to deal with a whole new sentence.

With a heavyweight list, as in this example, the list should be introduced by a colon; and the list members should be separated from each other by semi-colons. If you have a list that is lightweight, it should be introduced by a double-length hyphen (an "m-dash"); and the examples should be separated from each other by commas:

> I can get you the model 174B SUV in three colors — red, white, and blue.

The colon can also signal the reader that we will be going over the same ground again, either in different terms or from a different perspective:

> This will not be easy to accomplish: We are dealing with three different prospective partners who all have their own concerns that may interfere with ours.

The semi-colon, so underused in contemporary English, says something both subtle and clear to readers: "Dear reader, since this is just the end of a main clause, which could stand by itself as a sentence, please stress the words that come just before me; but hang on to that thought in your mind because it is the first part of a two-part thought. The second part will be starting immediately, stretching forward to the Stress Position created at the end of the sentence by the period. Thank you for your attention."

Scientists constantly deal with concepts or arguments that would benefit from a two-part structure, complete with two separate Stress Positions. In America, the semi-colon is rarely taught to students because at their young age they have not progressed far enough intellectually to have a constant need for two-part thoughts. When we do reach that age of intellectual development, no one is around to introduce us to its great value. Scientists need to get friendly with the semi-colon.

When the writer puts the emphatic material of a sentence in any place other than the Stress Position, one of two things can happen. Both are bad. First, the reader might find the Stress Position occupied by material that clearly is not worthy of emphasis. In that case, readers must discern for themselves what else in the sentence might be the most likely candidate for emphasis. There are no secondary structural indications to fall back upon. In sentences that are long, dense, or sophisticated, the alternative options are so numerous that readers are just left to guess at what the writer was trying to say. In that process, readers must expend far too much reader energy in their new, unwanted role of co-author. Long sentences are not long because they contain too many words; they are long because they contain too few Stress Positions. No wonder reading science is hard.

The second possibility is even worse: The reader may find the Stress Position occupied by something that does indeed appear capable of receiving emphasis, even though the writer did not intend it to be stressed. In that case, the reader is highly likely to emphasize this imposter material. The writer will have lost an important opportunity to influence the reader's interpretive process.

The Stress Position can change in size from sentence to sentence. It can vary from a single word to several lines. So here is another way of defining the Stress Position: A reader has reached the beginning of the Stress Position when it is apparent that there is nothing left in this sentence but the material presently being read. Thus, a whole list, numbered and indented, can occupy the Stress Position of a sentence, if it has been clearly announced as being all that remains of that sentence. Each member of that list, in turn, may have its own internal Stress Position, since each member may produce its own syntactic closure. A semi-colon at the end of each list member (with a period after the last one) will help make these emphases clear to a reader.

Putting the semi-colon to work, we can now devise a revision of Example 1's first sentence:

> The smallest of the URF's is URFA6L, a 207-nucleotide (nt) reading frame overlapping out of phase the NH_2-terminal portion of the adenosinetriphosphatase (ATPase) subunit 6 gene; it has been identified as the animal equivalent of the recently discovered yeast H^+-ATPase subunit 8 gene.

The reader is invited by the mid-sentence semicolon to stress the "sub-unit 6 gene"; and that, in turn, allows the reader to summon a new breath of reader energy with which to encounter the second of the sentence's two parts, with the "sub-unit 8 gene" receiving the necessary emphasis in the Stress Position created by the period.

Let me offer you both a warning and a comfort about learning to deposit stress-worthy material in a Stress Position: It is by no means as easy or mechanical a task as it first might sound. In fact, it is a very difficult thing to do — at first. It often takes a strenuous effort to identify which material deserves stress. That done, it is often difficult (again, at first) to figure out how to reconstruct the sentence's structure to get that material to a Stress Position. Why so difficult? Because you (like everyone else) are a creature of habit. In many ways, you are a prisoner of your habits. The first time you try to rewrite a document this way, you may fail to find a solution for as many as 50% of your sentences. But having endured the pain of the process, you will find yourself increasing that success percentage in every succeeding text. It will not take too many documents for this to become your new, natural way of constructing sentences.

It is important to begin this journey as a revision process. Do not try to write like this with a blank page in front of you until you have revised two or three documents successfully. It is a sharp learning curve, but does not last long. No single change you can make in your writing will bring you more success than this.

Let us now re-visit Example 1's awe-inspiring subject–verb separations in its 62-word long third sentence. I have bolded the grammatical subjects and underlined their verbs.

> Recently, however, immunoprecipitation **experiments** with antibodies to purified, rotenone-sensitive NADH-ubiquinone oxido-reductase [hereafter referred to as respiratory chain NADH dehydrogenase or complex I] from bovine heart, as well as enzyme fractionation studies, have indicated that **six human URF's** (that is, URF1, URF2, URF3, URF4, URF4L, and URF5, hereafter referred to as ND1, ND2, ND3, ND4, ND4L and ND5) encode subunits of complex I.

After encountering the subject ("experiments"), the reader must wade through 27 words (including three hyphenated compound words, a parenthetical interruption, and an "as well as" phrase) before alighting on the highly uninformative and disappointingly anticlimactic verb ("have indicated"). Without a moment to recover, the reader is handed a "that" clause in which the new subject ("six human URF's") is separated from its verb ("encode") by yet another 20 words. No wonder this prose fatigues us: too much waiting; too much energy expended on trying to remember the subjects while waiting for their verbs; too little sense of fulfillment when you arrive.

For Example 1, if ten editors were to remedy the subject–verb separations, and to find Stress Positions for what they considered the stress-worthy material, we would find that they had produced a number of differently revised texts — at least three or four, and maybe ten. Why not just one? If the writer's intentions had been crystal clear to all ten editors, they would all have come up with the same revised version. But each of them would have to be making best guesses at what the writer had intended; and that might produce as many different revisions as they had differing guesses. Example 1 is not merely "difficult"; it is absolutely "bad." It fails to communicate clearly to almost all readers what the writer was trying to say. That goes way beyond the calumnifying judgments of "long," "difficult," or "technical."

It is now commonplace in the world of language theory to accept that all sentences are infinitely interpretable, given an infinite number of interpreters. We cannot succeed in making even a single sentence mean one and only one thing; but we can increase the odds that a large majority of readers will tend to interpret our discourse according to our intentions. Such an increase will be fueled by writers becoming more consciously aware of how reader expectations control the interpretive process.

I never spoke with the author of Example 1. The revision I offer here, co-authored with Dr. Judith Swan back in 1990, is therefore the result of our best guesses:

> The smallest of the URF's, URFA6L, has been identified as the animal equivalent of the recently discovered yeast H^+-ATPase subunit 8 gene; but the functional significance of other URF's has been more elusive. Recently, however, several human URF's have been shown to encode subunits of rotenone-sensitive NADH-ubiquinone

> oxido-reductase. This is a large complex that also contains many subunits synthesized in the cytoplasm; it will be referred to hereafter as respiratory chain NADH dehydrogenase or complex I. Six subunits of Complex I were shown by enzyme fractionation studies and immunoprecipitation experiments to be encoded by six human URF's (URF1, URF2, URF3, URF4, URF4L, and URF5); these URF's will be referred to subsequently as ND1, ND2, ND3, ND4, ND4L and ND5.

Sheer length was not the problem; reducing length was not a pathway toward the solution. The revised version is not noticeably shorter than the original; but it is significantly easier to read and understand. We made our decisions not as an effort to decrease the number of words, but on the structural location of information, based on reader expectations. (See especially the last sentence of our revision.) If the author were to tell us this revision failed to grasp what she was trying to say, the fault would be not ours but hers. Her job was to say what had been on her mind — and to do it clearly.

When is a sentence too long? The creators of readability formulas would have us believe there exists some fixed number of words (the favorite is 29), past which a sentence is too hard to read. I disagree. I have seen ten-word sentences that are virtually impenetrable and 100-word sentences that flow effortlessly to their multiple points of resolution. In place of the word-limit concept, I offer the following definition: A sentence is too long when it has more viable candidates for Stress Positions than it has Stress Positions. Without the Stress Position's locational clue that its material is intended to be emphasized, readers are left too much to their own devices. They constantly have to assume the role of co-author, deciding what in the sentence ought to be considered most important. That extra effort is what produces the sensation of "too long." It calls for far too much reader energy. It is a major cause of reader fatigue.

In revising the example passage, we made decisions about what to omit and what to emphasize. We put subjects and verbs together to lessen the reader's syntactic burdens; we put the material we believed worthy of emphasis in Stress Positions; and we discarded material for which we could not discern significant connections. In doing so, we produced a clearer passage — but not one that necessarily reflected the author's intentions; it reflects only our interpretation of the author's intentions. The more problematic the structure, the less likely it becomes that a grand majority of readers will perceive the discourse in exactly the way the author intended. No one should have to work as hard as we did to unearth the content of a single passage of this length.

CHAPTER 4

THE TOPIC POSITION

In most sentences, the Topic Position includes everything from its beginning up to but not including the verb.

To summarize the principles connected with the Stress Position, we have the proverbial wisdom, "Save the best for last"; to summarize the principles connected with the topic position, we have its proverbial contradiction, "First things first." In the Stress Position, the reader needs and expects closure and fulfillment; in the topic position, the reader needs and expects perspective and context. With so much of reading comprehension affected by what shows up in the topic position, it behooves a writer to control the beginning of sentences with great care.

The information that begins a sentence establishes for the reader a perspective for viewing the sentence as a unit: Readers expect an English clause (a unit of discourse that contains a subject and a verb) to be the story about whoever or whatever shows up early on as the grammatical subject. "Bees disperse pollen" and "Pollen is dispersed by bees" are two different but equally functional sentences about the same fact. "Bees disperse pollen" tells us something about bees; "Pollen is dispersed by bees" tells us something about pollen.

If the grammatical subject tells us whose story this is, then the verb inescapably tells us what action that subject is doing: The subject "does" the verb; and the verb "is done by" the subject. (If the verb is in the passive mode, this changes: Then the verb "is done to" the subject.)

Any well written scientific sentence is likely to contain a number — often a large number — of important words, each one of which is capable of being linked to the following sentence. Starting that new sentence, readers have to know, unambiguously, which word or words in the previous sentence are to be carried forward as the logical connection to the sentence beginning to unfold. Since the reader is going to try to make this logical link with the first words encountered that are capable of making that link, it is our responsibility as writers to make sure that the appropriate linking words appear as early in the next sentence as possible.

As an example, here is a ten-word sentence that, by itself, seems to offer no difficulties:

> The overall scope is the greatest strength of the proposal.

A sentence that follows this one could logically carry on with any of the important-sounding words here — "overall," "scope," "greatest," "strength," or "proposal." A reader wants — indeed, needs — to know as soon as possible in the next sentence the word or words that will successfully make that logical connection. If the thought is intended to proceed with more discussion of the "scope," locating some reference to "scope" as soon as possible in the new sentence will accomplish that linkage. But if the word "strength" appears in the new sentence before anything connected to "scope" has a chance to appear, the logical link will be destroyed: Most readers will try to make this new sentence into a continuing story of "strength."

Equally difficult would be a new sentence that proceeds with no backward link whatever to the previous sentence. Then it is left to the reader (now once again transformed into a co-author) to figure out how these two sentences are meant to go together.

Putting the right piece of information in a Stress Position will clarify what that individual sentence is trying to say; but equally important is the transportation of the reader's mind from one sentence to another, so that the thought is allowed to develop from sentence to sentence. These two reader expectations — stress-worthy information in the Stress Position, and the appropriate backward linking information placed as soon in the new sentence as possible — are the two most important things to understand about how the English language functions. Do these well on a regular basis, and you will be a fine writer of English.

The topic position, at the sentence's beginning, must therefore convey two important things to a reader: (1) whose story this will be (located in the grammatical subject); and (2) how this new sentence should be connected to its predecessor. The action, located in the verb (which I will discuss later on), appears immediately thereafter. Taken together, these three combine to act as the context for this new sentence. Context controls meaning. Once that context is properly established, it is time for new information to appear and interact and build to the climax of what is to be most stressed. That is the shape of a well-written one-clause sentence in the English language.

Let us look at all three of these reader expectations functioning together.

Example 2:

Large earthquakes along a given fault segment do not occur at random intervals because it takes time to accumulate the strain energy for the rupture. The rates at which tectonic plates move and accumulate strain at their boundaries are approximately uniform. Therefore, in first approximation, one may expect that large ruptures of the same

> fault segment will occur at approximately constant time intervals. If subsequent main shocks have different amounts of slip across the fault, then the recurrence time may vary, and the basic idea of periodic mainshocks must be modified. For great plate boundary ruptures, the length and slip often vary by a factor of 2. Along the southern segment of the San Andreas fault, the recurrence interval is 145 years with variations of several decades. The smaller the standard deviation of the average recurrence interval, the more specific could be the long term prediction of a future mainshock.

This is the kind of passage that in subtle ways can make readers feel bad about themselves. The individual sentences give the impression of being intelligently fashioned: They are not especially long nor convoluted; their vocabulary, though appropriately professional, is not beyond the ken of most educated readers; and they are free of grammatical and dictional errors. On first reading, however, many of us arrive at the paragraph's end without a clear sense of where we have been or where we were supposed to have been going. When that happens, we tend to berate ourselves for not having paid close enough attention. In this case — and in so many cases in English professional prose — the fault lies not with us, but with the author.

We can start to uncover the problems by looking closely at the grammatical subjects of all the clauses, to learn what story each sentence is purporting to tell:

> Large earthquakes
> The rates
> [...] one
> if subsequent mainshocks [...] then the recurrence time
> the length and slip
> the recurrence interval
> the smaller [...] the more specific

Much of this information is making its first appearance in this paragraph. With constantly new material showing up as grammatical subjects, the focus of the story — the "whose story" — is constantly shifting. Given just the material in the topic positions, no two randomly selected readers would be likely to construct exactly the same story as the one the paragraph was intended to tell.

If we try to piece together the relationship of each sentence to its neighbors, we notice that certain bits of old information keep reappearing. We hear a good deal about the recurrence time between earthquakes: The first sentence introduces the concept of nonrandom intervals between earthquakes; the second sentence tells us that recurrence rates due to the movement of tectonic plates are more or less uniform; the third sentence adds that the recurrence rates of major earthquakes should also be somewhat predictable; the fourth sentence adds that recurrence rates vary with some conditions; the fifth sentence adds information about one particular variation; the sixth sentence adds a recurrence-rate example from California; and the last sentence tells us something about how recurrence rates can be described statistically. This refrain of "recurrence intervals" in fact constitutes the major string of repeated information in the paragraph. This paragraph is struggling to be the emerging story about those rates.

Unfortunately, the rates of recurrence intervals clearly appear in the "whose story" position only twice. The verbs of all the other sentences are telling us what is going on in a number of other stories — not the story of the rates. In addition, almost every sentence raises questions about how it connects to the one before it, since those connections are not made explicit at the beginning of each sentence. All three elements of the paragraph's contextual connections — whose story, what action, and helpful backward links — are not located where they could best do their jobs. In the sentences as we have them, we have to keep asking the important questions — whose story? backward link? greatest stress? — all the way through. We are asked to use much more reader energy than we would have been likely to have summoned. We are once again forced into playing the role of co-author. The threat of misinterpretation sets in, bringing with it fatigue.

Here is one possible revision: It relies on these guesses; it makes further guesses as to whose story is being told at any given moment; and it makes yet further guesses as to what the logical links might be. As readers, we should not have to do all this co-author kind of guessing.

Once again: If the editorial decisions made below, concerning the relative values of specific information here, differ from yours, we can all blame the author, who failed to make his intentions apparent.

Here is a list of what seemed to us to be the new, most stress-worthy material in each sentence:

> time to accumulate strain energy along a fault
> approximately uniform
> large ruptures of the same fault
> different amounts of slip
> vary by a factor of 2
> variations of several decades
> predictions of future mainshock

Relying on these guesses, and making further guesses as to whose story is being told at any given moment, and what the logical links might be, here is a possible revision:

> Large earthquakes along a given fault segment do not occur at random intervals because it takes time to accumulate the strain energy for the rupture. The rates at which tectonic plates move and accumulate strain at their boundaries are roughly uniform. Therefore, nearly constant time intervals (at first approximation) would be expected between large ruptures of the same fault segment. [However?], the recurrence time may vary; the

> basic idea of periodic mainshocks may need to be modified if subsequent mainshocks have different amounts of slip across the fault. [Indeed?], the length and slip of great plate boundary ruptures often vary by a factor of 2. [For example?], the recurrence intervals along the southern segment of the San Andreas fault is 145 years — but with variations of several decades. The smaller the standard deviation of the average recurrence interval, the more specific could be the long term prediction of a future mainshock.

We are making progress: All the sentences end with what might be most important. From the second sentence on, they are all the story of the rates and their intervals. But having clarified those two issues, we now discover that there are all sorts of problems in the original text that were lost in the general misplacement of thought:

— Is the reason earthquakes do not occur at random intervals stated in the first sentence or in the second?

— Are the suggested choices of "however," "indeed," and "for example" the right ones to express the connections at those points? (All these connections were left unarticulated in the original paragraph.)

— If "for example" is an inaccurate transitional phrase, then exactly how does the San Andreas fault example connect to ruptures that "vary by a factor of 2"?

> — Is the author arguing that recurrence rates must vary because fault movements often vary?
>
> Or is the author preparing us for a discussion of how in spite of such variance we might still be able to predict earthquakes? This last question remains unanswered because the final sentence leaves behind earthquakes that recur at variable intervals; it switches instead to earthquakes that recur regularly.
>
> Given that this is the first paragraph of the article, which type of earthquake will the article most likely proceed to discuss?

In sum, we are now aware of how much the paragraph had *not* communicated to us on first reading. We can see that most of our difficulty was owing not to any deficiency in our reading skills but rather to the author's lack of comprehension of our structural needs as readers. It may have "sounded like" a competent discussion was going forward; but all this lack of clarity did not become apparent until our attending to the structural needs uncovered the lack of logical connections.

REA is not just a way to clean up prose that appears cluttered; it is a structural method of entering deeper into the production and communication of substance. Do not think of REA as a set of new rules, like "Always put the most important thing in the Stress Position" — which suggests that all will be well when you have done that. Instead, REA is your invitation back into your thinking process. To answer the structural question "What words shall I move over into the Stress Position?," you first have to answer the more substantive question "What was I trying to say here that was important?" You cannot place the backward-linking old information up front until you know what connection you intend between the last sentence and this new one. Writing well leads to thinking well leads to writing well leads to [...]."

The two major problems in professional prose, especially in the sciences, are these:

> People most often fail to locate the stress-worthy information in a Stress Position; and
>
> People often fail to provide the reader with the necessary backward link at the beginning of every new sentence.

Why should these problems appear so constantly in the prose of very intelligent people? The source of the problem is not hard to discover: Most writers produce prose linearly (from left to right) and through time. As they begin to formulate a sentence, often their primary anxiety is to capture the important new thought in words before it escapes. Quite naturally, they rush to record that new information on paper, after which they can produce, at their leisure, contextualizing material that links back to the previous discourse. Writers who do this consistently are attending more to their own need for unburdening themselves of their information than to the reader's need for receiving it. REA explicitly explains where readers tend to look in your sentences for the answers to the crucial questions concerning context and emphasis.

Now that these concepts have been posited and somewhat explained, let us look at another example in which attending to what may seem "merely" structural matters brings us right to the heart of the substantive problems with the paragraph's thought.

When the correct backward linking old information does not appear anywhere in a sentence, readers are left to construct the logical linkage by themselves. Often this happens when the connections are so clear in the writer's mind that they seem unnecessary to state; at those moments, writers underestimate the difficulties and ambiguities inherent in the reading process. This third example attempts to demonstrate how paying close attention to the placement of contextual old information and emphatic new information can reveal where a writer has neglected to articulate essential connections.

Example 3:

The enthalpy of hydrogen bond formation between the nucleoside bases 2′deoxyguanosine (dG) and 2′deoxycytidine (dC) has been determined by direct measurement. dG and dC were derivatized at the 5′ and 3′ hydroxyls with triisopropylsilyl groups to obtain solubility of

> the nucleosides in non-aqueous solvents and to prevent the ribose hydroxyls from forming hydrogen bonds. From isoperibolic titration measurements, the enthalpy of dC:dG base pair formation is −6.65±0.32 kcal/mol.

Although part of the difficulty of reading this passage may stem from its abundance of specialized technical terms, a great deal more of the difficulty results from its structural problems. These problems are now familiar to us: We are not sure at all times whose story is being told; in the first sentence the subject and verb are widely separated; the second sentence has only one Stress Position but two or three pieces of information that are probably worthy of emphasis — "solubility [...] solvents," "prevent [...] from forming hydrogen bonds" and perhaps "triisopropylsilyl groups." These perceptions, acted upon in revision, produced the following improvement:

1. Invert the first sentence, so that (a) the subject-verb-complement connection is unbroken, and (b) "dG" and "dC" are introduced in the Stress Position as new and interesting information. (Note that inverting the sentence requires stating who made the measurement; since the authors performed the first direct measurement, recognizing their agency in the topic position may well be appropriate.)

2. Since "dG" and "dC" become the backward-linking old information in the second sentence, keep them up front in the topic position.

3. Since "triisopropylsilyl groups" is new and important information here, create for it a Stress Position.

4. "Triisopropylsilyl groups" then becomes the old information of the clause in which its effects are described; place it in the topic position of that clause, where it will function as the backward link and also become whose story the clause will be.

5. Alert the reader to expect the arrival of two distinct effects by using the flag word "both." "Both" notifies the reader that two pieces of new information are meant to be stressed in a single Stress Position.

Here is a partial revision based on these decisions:

> We have directly measured the enthalpy of hydrogen bond formation between the nucleoside bases 2′deoxyguanosine (dG) and 2′deoxycytidine (dC). dG and dC were derivatized at the 5′ and 3′ hydroxyls with triisopropylsilyl groups; these groups serve both to solubilize the nucleosides in non-aqueous solvents and to prevent the ribose hydroxyls from forming hydrogen bonds. From isoperibolic titration measurements, the enthalpy of dC:dG base pair formation is −6.65±0.32 kcal/mol.

Although the outlines of the experiment are now becoming visible, there is still a major logical gap. After reading the second sentence, we expect to hear more about the two effects that were important enough to merit placement in its Stress Position. Our expectations are frustrated, however, when those effects are not mentioned in the next sentence: "From isoperibolic titration measurements, the enthalpy of dC:dG base pair formation is −6.65±0.32 kcal/mol." The authors have neglected to explain the relationship between the derivatization they performed (in the second sentence) and the measurements they made (in the third sentence). Ironically, that is the point they probably most wished to make here.

At this juncture, particularly astute readers who are chemists might draw upon their specialized knowledge, silently supplying the missing connection. Other readers are left in the dark. Here is one version of what the authors seem to mean to say, with two additional sentences supplied from Dr. Swan's knowledge of nucleic acid chemistry:

> We have directly measured the enthalpy of hydrogen bond formation between the nucleoside bases 2′deoxyguanosine (dG) and 2′deoxycytidine (dC). dG and dC were derivatized at the 5′ and 3′ hydroxyls with triisopropylsilyl groups; these groups serve both to solubilize the nucleosides in non-aqueous solvents and to prevent the ribose hydroxyls from forming hydrogen bonds. Consequently, when the derivatized nucleosides

> are dissolved in non-aqueous solvents, hydrogen bonds form almost exclusively between the bases. Since the interbase hydrogen bonds are the only bonds to form upon mixing, their enthalpy of formation can be determined directly by measuring the enthalpy of mixing. From our isoperibolic titration measurements, the enthalpy of dG:dC base pair formation is −6.65±0.32 kcal/mol.

Each sentence now proceeds logically from its predecessor. We never have to wander too far into a sentence without being told where we are and what former strands of discourse are being continued. And the "measurements" of the last sentence have now become old information, reaching back to the "measured directly" of the preceding sentence. (It also fulfills the promise of the "we have directly measured" with which the paragraph began.) This was all accomplished by attending to the placement of information according to the principles of REA.

CHAPTER 5

VERBS AS ACTIONS; ACTIONS AS VERBS

I have mentioned, almost in passing, that readers of English expect that the action taking place in a clause will be located in its verb. Once the subject has articulated whose story this is, we lean forward to the verb to discover what action is taking place. This concept deserves an example of its own.

Example 4:

Transcription of the 5*S* RNA genes in the egg extract is TFIIIA-dependent. This is surprising, because the concentration of TFIIIA is the same as in the oocyte nuclear extract. The other transcription factors and RNA polymerase III are presumed to be in excess over available TFIIIA, because tRNA genes are transcribed in the egg extract. The addition of egg extract to the oocyte nuclear extract has two effects on transcription efficiency. First, there is a general inhibition of transcription that can be alleviated in part by supplementation with high concentrations

> of RNA polymerase III. Second, egg extract destabilizes transcription complexes formed with oocyte but not somatic 5*S* RNA genes.

There are all sorts of problems here. If we are to concentrate for a moment on verbs articulating actions, why produce an example that has so many other problems as well? The answer: When a sentence has a serious structural problem concerning one reader expectation, there is a high likelihood that this one problem will generate a series of other problems. That makes sense: If X is not where the reader needs it to be, then it may wind up where the reader needed Y to be [...], and so forth. It can create a structural domino effect. Given this, it does not matter which structural problem you attend to first, since that one will set in motion a complete overhaul of the sentence's structure. So in dealing with Example 4, we should quickly check whose story each sentence probably is, on our way to contemplating what actions are happening. The two are inseparably linked.

When we look at the grammatical subjects in each of these sentences, we discover that we cannot tell whose story the passage is: The story's focus changes in every sentence. If we search for repeated old information in hope of settling on a good candidate for several of the topic positions, we find all too much of it — egg extract, TFIIIA, oocyte extract, RNA polymerase III, 5*S* RNA, and transcription. All of these reappear at various points; but none announces itself clearly as our primary focus. It appears that the passage is trying to tell several stories simultaneously, allowing none to dominate. This may well be not just our problem as editors; the author may have been as equally uncertain as we are. Let us look at the Example's verbs, to see if they help us to understand what actions are taking place:

is
is [...] is
are presumed to be
are transcribed
has
is [...] can be alleviated
destabilizes

The list gives us too few clues as to what actions actually are supposed to be taking place. We as readers are again constrained to act as co-author. If the actions are not to be found in the verbs, we as readers have no secondary structural clues for where to locate them. Each of us has to make our personal interpretive guesses; the writer has failed to control the reader's interpretive act.

Worse still, in this passage some of the most important actions never actually appear. Based on a reasonable and informed guess, the verbs that connect these players seem to be "limit" and "inhibit." If we express those actions as verbs, and place the most frequently occurring information — "egg extract" and "TFIIIA" — in the topic position whenever possible, we generate the following revision:

> In the egg extract, the availability of TFIIIA limits transcription of the 5*S* RNA genes. This is surprising because the same concentration of TFIIIA does not limit transcription in the oocyte nuclear extract. In the egg extract, transcription is not limited by RNA polymerase or other factors because transcription of tRNA genes indicates that these factors are in excess over available TFIIIA. When added to the nuclear extract, the egg extract affected the efficiency of transcription in two ways. First, it inhibited transcription generally; this inhibition could be alleviated in part by supplementing the mixture with high concentrations of RNA polymerase III. Second, the egg extract destabilized transcription complexes formed by oocyte but not by somatic 5*S* genes.

The choice of "limit" and "inhibit" around which to organize the revision is only one possible guess. Might you have chosen others? Surely. All readers make exactly that kind of choice in the reading of every sentence. The fewer the structural clues to interpretation given by the author, the more variable the resulting interpretations will tend to be. If you and I were to argue about differences in interpretations, we should remember — once again — that this is the fault of neither you nor me: It is the fault of the writer. She should have resolved the questions for us by articulating those actions as verbs. There should be no questions, no need to guess.

As a story about "egg extract," this passage still leaves something to be desired. But at least now we can recognize that the author has not explained the connection between "limit" and "inhibit." This unarticulated connection seems to contain both of her hypotheses: First, that the limitation on transcription is caused by an inhibitor of TFIIIA present in the egg extract; and, second, that the action of that inhibitor can be detected both by adding the egg extract to the oocyte extract and then by examining the effects on transcription. As critical scientific readers, we would like to concentrate our energy on whether the experiments prove the hypotheses. We cannot begin to do so if we are left in doubt as to what those hypotheses might be. We cannot begin to do so if we are using most of our reader energy to make sense of the structure of the prose; the structure was supposed to help us make sense of the substance.

CHAPTER 6

ONE STRUCTURAL WEAKNESS THAT AFFLICTS ALMOST ALL WRITERS IN ENGLISH

We have been exploring how the structural location of information in a sentence acts as a kind of instructional manual that helps the reader to discover the ideas being communicated by a writer. I wish to add to this a second, smaller, but highly significant instructional manual, this time stemming from the nature of three grammatical units of discourse. The three units are (1) the main clause, (2) the qualifying clause, and (3) the phrase. Each of these, by its very nature, sends an interpretive instruction to the reader. No one to date has recognized a major reading problem that is constantly being created by our never having been taught what these units of discourse are telling us.

In English, a clause is any unit of discourse that contains both a subject and a verb. There are two kinds of clauses — the "main clause," and what I am calling the "qualifying clause." (The latter has always been referred to as a "dependent clause"; since I believe that term to be misleading, I am renaming it "qualifying clause," which better describes its function.)

A "main clause" contains a subject and a verb and can stand by itself as a complete sentence. You can put a capital letter at its beginning and a period at its end, and no grammar teacher could deduct any points for error. As such, it sends the following subtle, silent message to the reader: "Dear reader, I'm a main clause and could stand by myself as a sentence. I therefore must contain something worthy of your emphasis. Please stress something in me."

A "qualifying clause" contains a subject and a verb but cannot stand by itself as a complete sentence. If you begin it with a capital letter and put a period at its end, your grammar teacher would write "sentence fragment" in the margin and deduct points. As such, it sends the following subtle, silent message to the reader: "Dear reader, being a qualifying clause, and having my own subject and verb, I am elegant, significant, and worthy of your attention; my job is not to convey what is most important, but rather to 'qualify' my older sibling, the main clause. So please do not stress anything in me."

Those two together cover all clauses. The only other category that is important for us to know and understand is the "phrase" – a unit of discourse with a beginning and an end that does not include both a subject and a verb. These three categories of units of discourse cover everything. The phrase sends the following subtle, silent message to the reader: "Dear reader, since I do not contain a subject and verb, I am a much lighter weight unit of discourse then either of the two clauses. So do not bother emphasizing anything that I have to say. Do not stress anything in me."

A large percentage of Americans were never taught grammar, as a result of a series of studies in the 1970s that argued there was no connection between the knowledge of grammar and the quality of writing. Those who were taught grammar were asked to memorize it as rules. I remember learning all about the main clause, the independent clause, the dependent clause, the compound clause, the complex clause, and (my personal favorite) the compound-complex clause. The moment I turned in my grammar exam paper on these matters, all these distinctions evaporated from my mind — and rightfully so.

No one had taught me why these distinctions were important or helpful. I came to realize, over decades of study, that we need to know only three of these units by name and by function — because, together, they are important and helpful: They send three different instructions to the reader. Since almost no one understands how these discourse units function, the writing problem that stems from this ignorance affects almost everyone. I am calling the problem "the main clause first problem." In the 16 years to date that I have been actively searching for it in students' and my clients' prose — and that includes the work of many thousands of individuals — I have yet to find a single person who does not suffer from this writing disease. It is pandemic; no one knows it even exists; and the cure for it is straightforward. I shall explain.

In the structural instructions we were exploring earlier, it was the Stress Position that announced what information was most worthy of stress. But now we must add to that a second, grammatical instruction on the same matter: The main clause tells us to stress something in it. The main clause first problem arises when these two instructions conflict with each other. Thus, any main clause that ends without a Stress Position (created by a colon, semi-colon, or period) creates a potential conflict for the reader: The main clause instructs the reader to stress something; but the lack of Stress Position punctuation tells the reader *not* to stress anything. This conflict forces the reader again to assume the role of co-author.

This is a major problem, especially for scientific writing. Picture a reader coming to the end of a main clause in the middle of a sentence with only a comma to be of help. At that moment, a small part of the reader's brain must direct its attention backwards, as the reader tries to decide whether or not to stress something in the first clause — and if so what. But most of the reader's brain will be hurtling forward due to the pressing need of finishing the sentence. Moments like this, when the brain is simultaneously trying to go in two opposite directions at once, cause confusion, cause misinterpretation, and cause an ever growing sense of fatigue.

How often does this main clause first problem occur in scientific writing? (It can only happen in a sentence that includes more than just a single main clause, since any such one-clause sentence must end with a period.) For many years, I used to guess that about 50% of scientific sentences (in grant proposals, published articles, and other professional documents) contained information that extended beyond the limits of a single main clause. I also used to guess that, of those 50% longer sentences, about 70% were negatively affected by the main clause first problem. In 2022, in preparation for publishing an article about this problem, I undertook the task of putting my guesses on trial. I chose five published articles from *Science Magazine*, all of which were about 4,500 words in length, and none of which had a first author's name that might suggest that English was not that person's first language. I did the laborious statistics. My first guess turned out to be right on: 49.8% of all the sentences extended themselves beyond a single main clause. My second guess, however, turned out to be way off: of those 49.8%, the main clause first problem was created not 70% of the time but rather 91% of the time. Ninety-one percent of 49.8% is 45.3%. That is the minimum percentage of scientific sentences in which a scientific reader is called upon to morph into the role of a scientific co-author. No wonder, as we conclude the reading of a scientific text, we feel that distinct sense of fatigue. More important, that is how often our understanding of a scientist's meaning stems primarily from our guesses, instead of the author's instructions. Once you learn to recognize the main clause first problem by sight, you are highly likely to find it in your prose. To overcome it, you must become able to recognize a main clause, a qualifying clause, and a phrase when you see them. You must come to be aware of when a main clause is ending. To help you on your way, here is what I hope will be an instructive example, the fifth and last.

Example 5:

> Inhibition of CMA in heart unexpectedly confers resistance to stress-induced cardiac dysfunction in both pressure overload and myocardial infarction models, which is opposite to MA deficiency in heart, and differs from CMA deficiency in other tissues.

This sentence could have ended at the first comma, without offending any grammar teacher, since those 20 words form a main clause plus a modification. The rest of the sentence is a 16-word qualifying clause — a "which" clause that cannot stand by itself as a sentence. When we reach the end of the main clause, and find no Stress Position, we are left on our own to decide if we should have stressed something in those first 20 words; but most of our attention is devoted to finishing the sentence. This is the main clause first problem.

However, the trouble starts earlier than at that comma. As a first-time reader, when I reach the 12th word, "dysfunction," I have already experienced enough for a whole main clause — and therefore have been handed enough for a whole sentence. Words 13–20 are a two-part phrase that further describes where the dysfunction might occur. The natural weight of that 12-word main clause suggests that something already encountered deserves emphasis. Again, if there had been a period there, a grammar teacher could not have taken off any points.

> Inhibition of CMA in heart unexpectedly confers resistance to stress-induced cardiac dysfunction //

But since, as a reader, I *could* have stopped there, my forward motion toward an expected closure of another Stress Position produces an interpretive tension in me. I could stop; I perhaps want to stop; but I am not allowed to stop. It is simultaneously both a retroactive-leaning and a forward-looking problem. It is the main clause first problem.

Retroactively, the question becomes how should I have been reading these 12 words? Since there is enough for a full sentence here, should I already have stressed something? And if so, what? Consider the number of reasonable candidates for stress: "inhibition of CMA"; "in heart"; "unexpectedly"; "confers resistance"; and "stress-induced cardiac dysfunction." That makes five. But perhaps should I be stressing more than one? Or, importantly, believing the lack of punctuation here, which tells me not to stress anything yet, should I perhaps be stressing nothing? All of these decisions have to be made by the reader retroactively at the clause's end; but at that confusing and unconfident moment, most of the mind is barreling forward to finish reading the sentence. We have been given two conflicting instructions at the 12-word mark — "stress something from the main clause here" and "do not stress anything here, since there is no Stress Position punctuation." Then the problem happens again, but in reverse: Throughout the rest of the sentence, we will be reading units that are *not* main clauses, and therefore tell us "do not stress anything here"; but at their end, we find a period that instructs us "stress something here." This is a serious reading problem — especially because it is highly likely to occur several times on every page.

It was the responsibility of the writer first to make those decisions for the reader and then to communicate them clearly. The sentence's structure, signaled by its punctuation, should allow the reader to continue reading forward without either mental interruption or undue interpretive burden. As readers, we should be able to trust that the most important material will appear in a Stress Position. If, as a writer, you can manage to do that for your readers on a regular basis, they will quickly learn to trust you. You will also free yourself of the main clause first problem.

It is worth repeating: In 16 years of investigating this epidemic problem, I have found not a single writer who is free from it.

So far, even though we have looked at only 12 words of our 36-word sentence example, we have already overtaxed our supply of reader energy for the whole sentence. The missing grammatical expectation and misleading instructions force us to make a great many guesses. We are yet again in the co-author business.

I reprint the example here with double-slashes at every moment the sentence might grammatically have come to an end.

> Inhibition of CMA in heart unexpectedly confers resistance to stress-induced cardiac dysfunction // in both pressure overload and myocardial infarction models, // which is opposite to MA deficiency in heart, // and differs from CMA deficiency in other tissues. //

Again, a sentence is "too long" not when it exceeds some arbitrarily chosen number of words but rather when it has more viable candidates for Stress Positions than it has Stress Positions.

At each of those double-slash moments, the reader once again has to decide what, if anything, should be stressed along the way. Recall that there were five reasonable candidates in the first 12 words alone. The sentence sounds professional and dignified; but it is an interpretive headache.

This sentence was written by a highly intelligent and professionally competent scientist; but if we try hard to decide for ourselves as readers which pieces of information we should stress here, we come away with a mistaken sense that she is somewhat disorganized, or perhaps merely a pedestrian thinker, or perhaps a hopelessly complex one. Over multiple occurrences of this on every page of the document, the cumulative effect is hard for the reader to bear. And, to make it much worse, once you have guessed wrong as to what the writer meant in one clause, all your subsequent guesses might be based more on sand than on concrete.

Here is the revision the author and I devised, based on her decisions concerning what deserved stress. We made sure not only that each main clause contained stress-worthy information, but that stress-worthy material appeared at the end of every main clause, in Stress Positions created by a colon, a semi-colon, and couple of periods. That eliminated the "stress something" versus "stress nothing" conflict. See if this is easier and clearer for you to read:

> Inhibition of CMA in heart unexpectedly confers resistance to stress-induced cardiac dysfunction. This happens in both pressure overload and myocardial infarction models: In heart, this is opposite to MA deficiency; in other tissues, it differs from CMA deficiency.

As readers of this, we can feel confident in leaning forward to the moments of stress signaled by the colon, the semi-colon, and the periods. Having perceived, clearly and easily, what she wanted us to stress, we are ready to move on to the next sentence. We are not fatigued. We are her readers, not her co-authors.

The main clause first problem: Everyone suffers from it; almost no one yet knows it exists; and yet here is the cure. Create a Stress Position for everything you want your reader to stress. That solves the problem.

I implore you once again not to turn every reader expectation into a hard and fast rule. Do not think that if you have ended a main clause with something other than a colon, semi-colon, or period that you have necessarily created a problem for your reader. In English, there is one particular kind of sentence that commonly begins with the main clause, followed by a comma, without creating a main clause first problem. It is a sentence that follows the comma with the word "because." In many such cases, the opening clause requires the weight and dignity of being a main clause because it describes something for which the causes will immediately be announced. Anything less than a main clause would not have produced that necessary weight and dignity; and yet the single thing this sentence wants to emphasize the most is the cause of it all. Any writing decision calls for judgment. REA gives you a basis for making such judgments.

When should you *not* end a sentence with the most stress-worthy material? When its powerful position there will unduly burden a reader who you know already understands what the most important words are. When should you *not* let the verb articulate the action? When articulating that action so straightforwardly might be burdensome, painful, or disrespectful of a particular reader. Good writing continually calls for good judgment. That is why the single best rule is "No Rules." I am not invoking chaos. The rules of grammar are rules, mostly to be followed carefully. But the principles of REA cannot be followed slavishly without their backfiring on occasion. Every reader expectation can be violated to good effect. Our best writers are also our most skillful violators.

Why is the main clause first problem so pandemic? I think I have discovered the reason. When we wrote anything as very young students, most of our sentences contained only one clause. As a result, our sentences averaged no more than 12 to 15 words – the average length of a single clause. Since our sentences were usually only one clause long, we began the act of writing a sentence by beginning the act of writing a main clause: They were one and the same thing for us. We developed a habit of beginning the sentence-writing process by beginning to write a main clause.

During our higher education, we developed the need to write longer, more structurally complex sentences; but there was no one around to tell us that a more complicated sentence structure brings with it a need to understand how multi-clause sentences function for readers. So as scientists, we average 26-28 words per sentence, with no sense whatsoever of whether we are writing main clauses or qualifying clauses or phrases. No one taught us anything about those units of discourse; and no one taught us anything about the reader's need for Stress Positions. No wonder absolutely everyone produces sentences that begin with a main clause with no Stress Position. When we finish writing that opening main clause, it makes sense to us, as writers, to put a comma and keep on going. Who knew that often created chaos for readers?

To make progress where this is concerned, therefore, we have to teach ourselves to become aware of what kind of unit of discourse we are writing at the moment – a main clause, a qualifying clause, or a phrase. When we are writing a main clause, we have to understand that a reader will want to stress something in it; in order for them to stress the right thing, we writers have to signal them with a well-occupied Stress Position. This is a major problem in scientific communication; and we now have a way to solve the problem. The general principle? (Not a rule, just a principle.) Give your readers a Stress Position for everything you want them especially to stress.

Why do so many scientists (and, truth be told, professionals in every field,) write such badly structured, fatiguing prose? I think it has something to do with the habit we got into when we were learning to write as teenagers. Then our audience was our English teacher. She or he seemed to know everything about whatever topic had been assigned. Our job, as we perceived it, was to *demonstrate* what good students we were by assembling as much information as possible and somehow depositing it on pages that were error-free and increasingly intelligent sounding. Once the information was down on the page, teacher would know what to do with it. We were certain of that. But when you became a practicing professional, there was no one at the graduation ceremony to tell you that your rhetorical task was about to change. You would no longer be producing prose to *demonstrate* to an expert that you had done enough work to be rewarded with a good grade. You would no longer be rewarded just for working hard or improving since the last time. By becoming a professional, you have become the expert. When you are writing a grant proposal, an essay for publication, or even a job application, your rhetorical task is no longer *demonstration*; it has become *communication*. You should no longer depend on your readers to figure out what you have not yet clarified for yourself. You run a real risk of failure if you leave the interpretive process entirely up to your readers. It is now your job to signal them, sentence after sentence, how you, the knowledgeable one, want them to put your words together to form meaning.

CHAPTER 7

WRITING AND THE SCIENTIFIC PROCESS

I began this essay by arguing that complex thoughts expressed in impenetrable prose can be rendered accessible and clear without minimizing any of their complexity. The ensuing examples of scientific writing have ranged from the merely cloudy to the virtually opaque; yet all of them could be made significantly more comprehensible by observing the REA structural principles that reveal how readers actually go about reading. These will help you ensure that the relative emphases of the substance coincide with the relative expectations for emphasis raised by the structure. Your signaling to your readers can be crystal clear.

The substance of science comprises more than the discovery and recording of data; it extends crucially to include the act of interpretation. It may seem obvious that a scientific document is incomplete without the interpretation of the writer; it may not be so obvious that the document cannot "exist" without the interpretation of each reader. Writers cannot "merely" record data, even if they try. In any recording or articulation, no matter how haphazard or confused, each word resides in one or more distinct structural locations. The resulting structure, even more than the meanings of individual words, significantly influences the reader during the act of interpretation. The question then becomes whether the structure created by the writer helps or hinders the reader in the process of interpreting the text. Once you as a writer become consciously aware of the clues all readers of English instinctively take from its structure, you can control both your writing process and their interpretation processes to such an extent that it will lead you to improve your thinking process. That is what REA can do for you.

****

For further writings and videos about REA, go to GeorgeGopen.com.

****

George David Gopen
Professor Emeritus of the Practice of Rhetoric
Duke University

www.ingramcontent.com/pod-product-compliance
Lightning Source LLC
LaVergne TN
LVHW100818030926
847095LV00028B/204
* 9 7 8 1 8 0 1 3 6 0 9 3 7 *